FUNGI

CHARLES ROTTER

CREATIVE EDUCATION

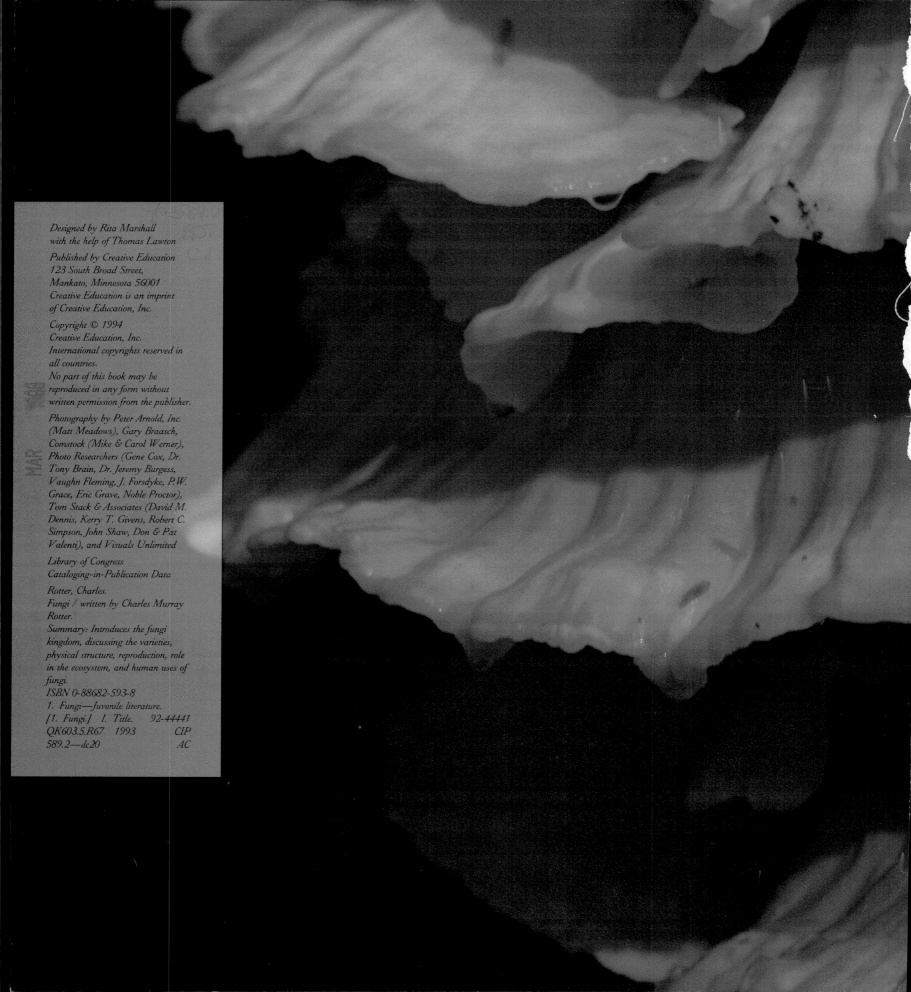

Designed by Rita Marshall
with the help of Thomas Lawton

Published by Creative Education
123 South Broad Street,
Mankato, Minnesota 56001
Creative Education is an imprint
of Creative Education, Inc.

Photography by Peter Arnold, Inc.
(Matt Meadows), Gary Braasch,
Comstock (Mike & Carol Werner),
Photo Researchers (Gene Cox, Dr.
Tony Brain, Dr. Jeremy Burgess,
Vaughn Fleming, J. Forsdyke, P.W.
Grace, Eric Grave, Noble Proctor),
Tom Stack & Associates (David M.
Dennis, Kerry T. Givens, Robert C.
Simpson, John Shaw, Don & Pat
Valenti), and Visuals Unlimited

Library of Congress
Cataloging-in-Publication Data

Rotter, Charles.
Fungi / written by Charles Murray
Rotter.
Summary: Introduces the fungi
kingdom, discussing the varieties,
physical structure, reproduction, role
in the ecosystem, and human uses of
fungi.
ISBN 0-88682-593-8
1. Fungi—Juvenile literature.
[1. Fungi.]   I. Title.   92-44441
QK603.5.R67   1993   CIP
589.2—dc20   AC

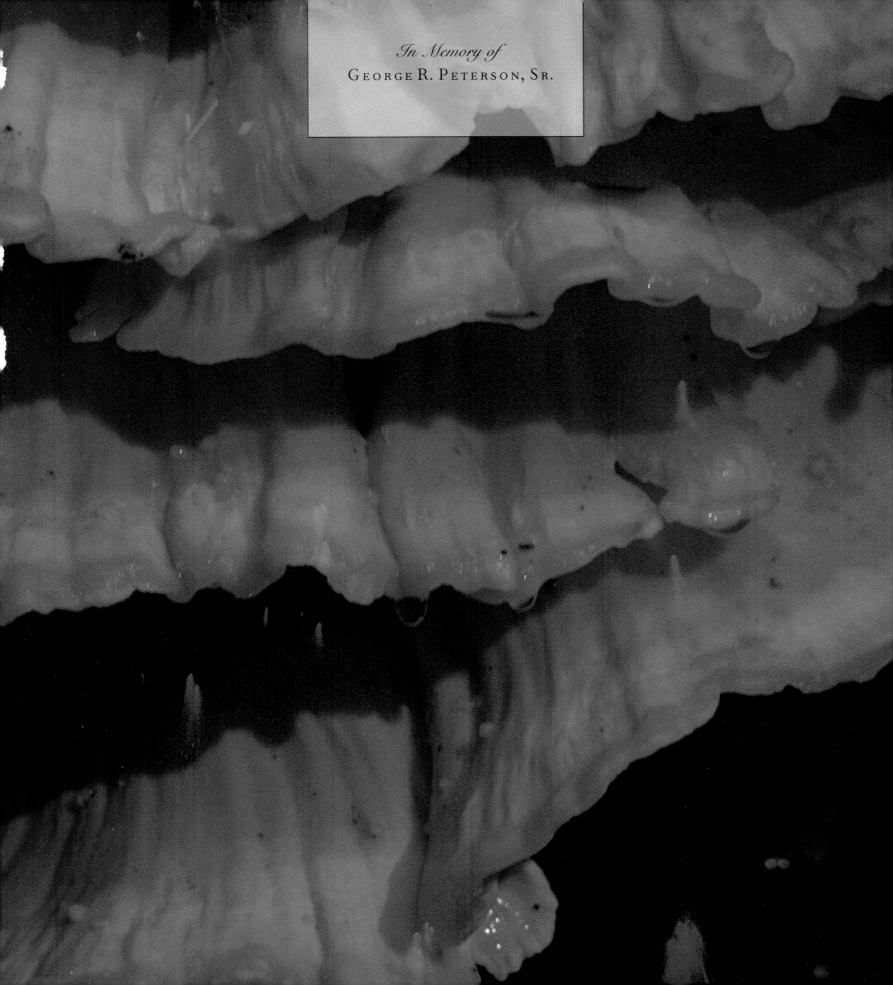

6

If you've ever seen mushrooms appear suddenly after a night of rain, or noticed a fuzzy mold spread quickly and engulf a piece of cheese, you've witnessed the sudden growth of a fungus. Fungi typically grow in cold, damp, or smelly locations. Associated with rot and decay, they are often considered a nuisance. But in fact, fungi play many essential roles in the world's ecosystem. From breaking down rocks into soil to providing life-saving medicines, fungi have a great impact on our life on earth.

Mushrooms push through the forest floor.

Until recently, fungi were thought of as primitive early plants, lacking the more developed features of modern plants, such as leaves or flowers. During the past century, however, new technologies have greatly increased our knowledge of living things. Researchers have been able to study the structure and even the chemical processes of living cells. Their discoveries have helped them realize that fungi are as far removed from plants as plants are from animals.

Rhizopus stolonifer, *or black bread mold.*

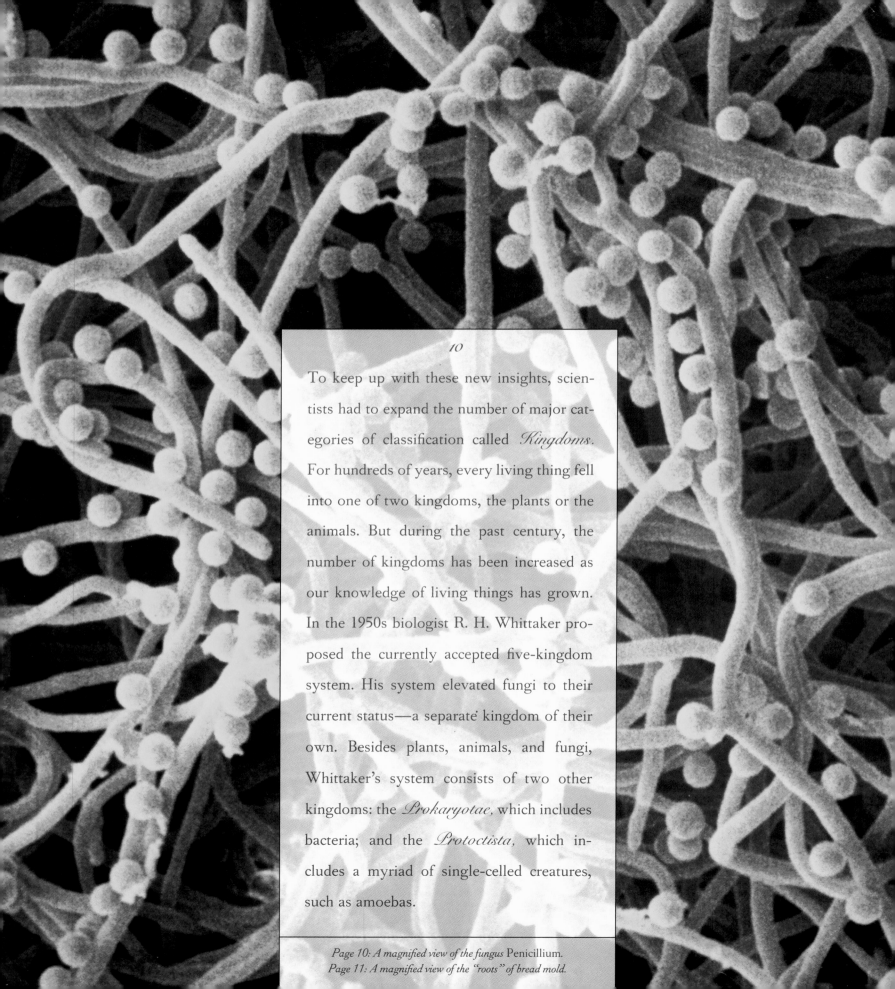

To keep up with these new insights, scientists had to expand the number of major categories of classification called *Kingdoms*. For hundreds of years, every living thing fell into one of two kingdoms, the plants or the animals. But during the past century, the number of kingdoms has been increased as our knowledge of living things has grown. In the 1950s biologist R. H. Whittaker proposed the currently accepted five-kingdom system. His system elevated fungi to their current status—a separate kingdom of their own. Besides plants, animals, and fungi, Whittaker's system consists of two other kingdoms: the *Prokaryotae,* which includes bacteria; and the *Protoctista,* which includes a myriad of single-celled creatures, such as amoebas.

Page 10: A magnified view of the fungus Penicillium.
Page 11: A magnified view of the "roots" of bread mold.

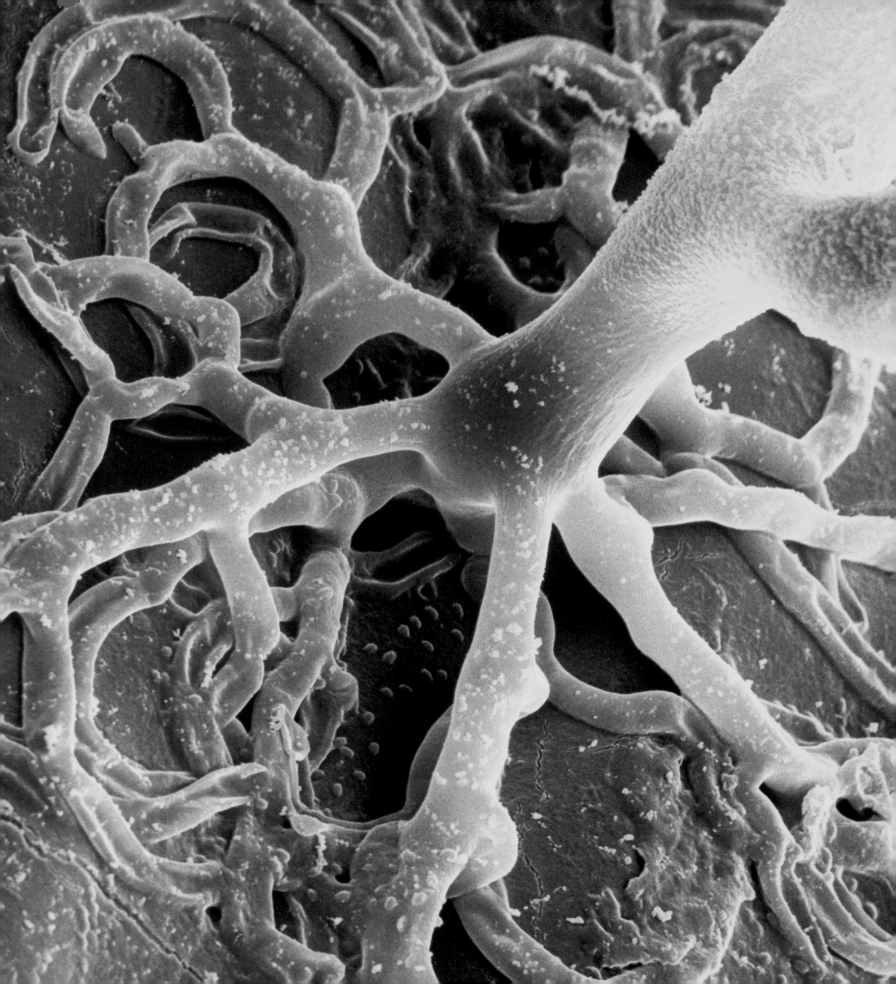

No other kingdom contains members with a wider spectrum of size than the fungi. Fungi range in size from the yeasts, one of the tiniest of living things, to a single fungus in the state of Washington that is the largest known living thing on earth. First publicly reported in May 1992, this single specimen covers 1,500 acres (608 hectares). The fungus, whose scientific name is *Armillaria ostoyae,* feeds on tree stumps. It grows mostly below the ground where it is not easy to see. The only visible parts of the fungus are the mushrooms that it produces. Scientists determined its size by taking tissue samples at different locations. Their analysis proved that the samples all came from one organism.

Hygrophorus *mushrooms.*

In nature, food is the chemical energy an organism needs to live. Fungi obtain their food differently from the way plants and animals do. Most plants, for instance, need sunlight to grow. In a process called *Photosynthesis*, plants use the sun's energy to convert nutrients from the air, water, and soil into food. Photosynthesis is the initial source of most of the food energy on earth.

Animals, on the other hand, usually obtain their food by ingesting, or eating, plants or other animals. After eating a food source, animals then digest the food, breaking it down into chemical energy inside their bodies. The energy that originated in the plants is transferred to animals by eating.

Raspberry slime mold.

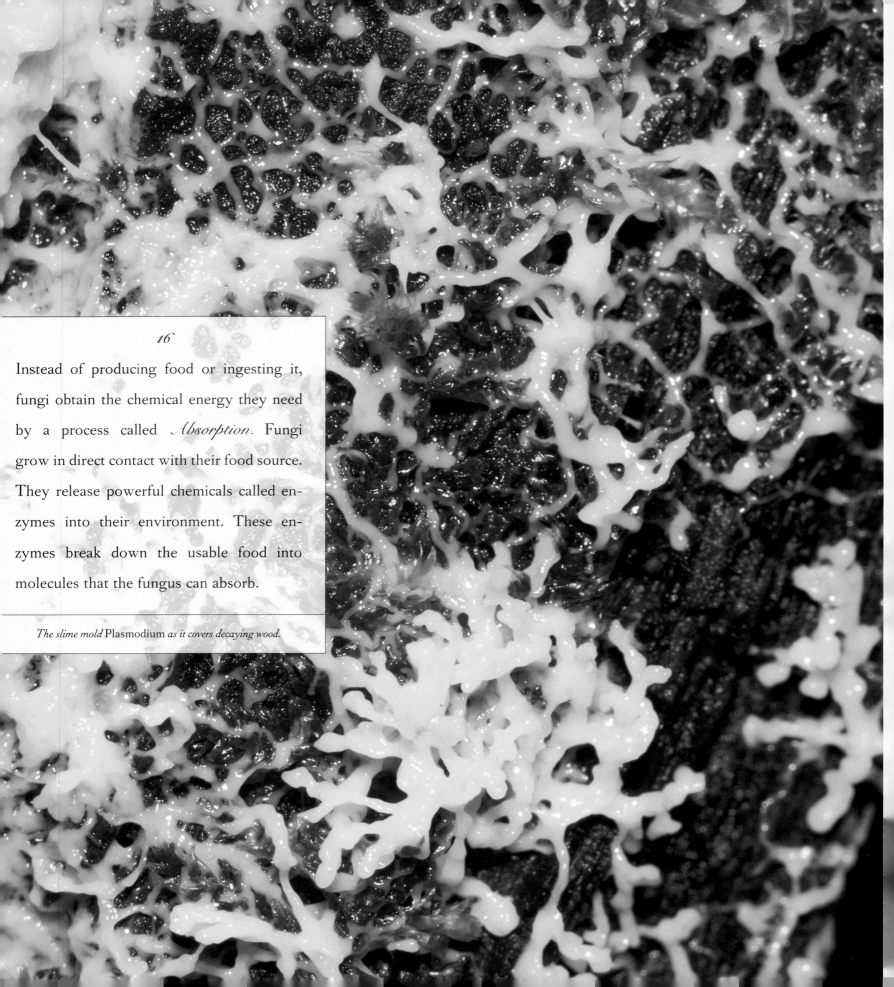

16

Instead of producing food or ingesting it, fungi obtain the chemical energy they need by a process called *Absorption*. Fungi grow in direct contact with their food source. They release powerful chemicals called en-zymes into their environment. These en-zymes break down the usable food into molecules that the fungus can absorb.

The slime mold Plasmodium *as it covers decaying wood.*

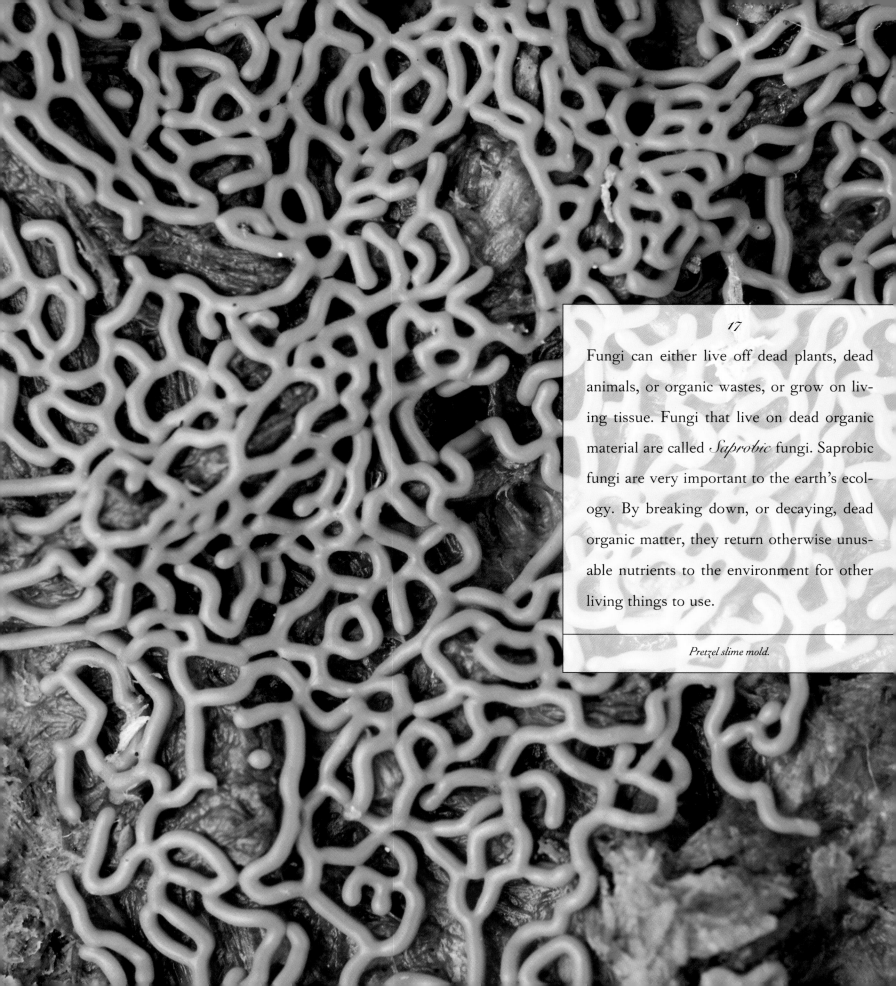

Fungi can either live off dead plants, dead animals, or organic wastes, or grow on living tissue. Fungi that live on dead organic material are called *Saprobic* fungi. Saprobic fungi are very important to the earth's ecology. By breaking down, or decaying, dead organic matter, they return otherwise unusable nutrients to the environment for other living things to use.

Pretzel slime mold.

When a fungus obtains its nourishment from living tissue, it is called a *Parasitic* fungus. Parasitic fungi attack living things as small as amoebas or as large as redwood trees. Parasitic fungi cause many diseases in both plants and animals. For example, ergot, a fungal disease of rye flowers, can cause serious illness in people who eat plants infected with it. The fungus produces toxins that can cause hallucinations and even death.

Different species that live together in an arrangement that is beneficial to both have what is called a *Symbiotic* relationship. Many fungi live in the roots of trees and other plants, helping the plants to obtain nutrients while receiving nourishment from them. Some scientists believe that symbiotic relationships between plants and fungi helped plants evolve into forms that could leave the water and spread out onto the land.

Clavulina amethystina.

Like all living things, fungi are made of cells. In most types of fungi, the cells join together to form slender tubes called *Hyphae* (singular hypha). The hyphae grow, branch, and spread out into the environment in search of nourishment. Often the hyphae cluster into a mass called a *Mycelium*. The mycelium normally appears as fuzzy threads; it is the main body of the fungus and is usually hidden underground.

Mold overtakes a piece of bread.
Inset: The fuzzy threads of a mycelium.

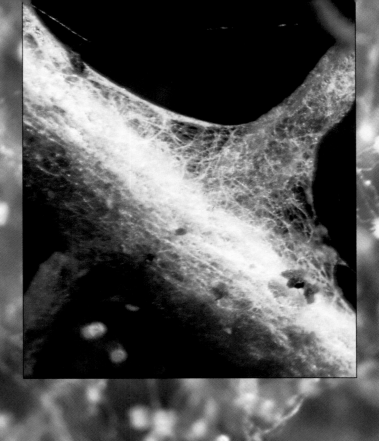

Hyphae can also develop into *Fruiting Bodies*, which enable the fungus to reproduce itself. Within the fruiting bodies of many fungi are special structures called *Sporangia* (singular sporangium) that manufacture spores. Like the seeds of flowering plants, spores may be dispersed in the wind or carried by insects to new locations. But unlike a seed, a *Spore* is a very simple structure, sometimes consisting of only a single cell. The spore starts to grow into a hypha when conditions are right. This hypha will then grow and branch into a new fungus.

22

Mushrooms are the most well-known fruiting bodies found in the fungi kingdom, but there are many others. Coral fungi, shelf fungi, truffles, puffballs, even the fuzzy surface of a mold—these are all fruiting bodies that produce spores by the millions, billions, or even more. A single giant puffball, which can grow up to 5 feet (1.5 m) across, can produce up to several trillion spores!

Page 22: A fruiting body growing out of a log.
Page 23: A puffball releasing spores.

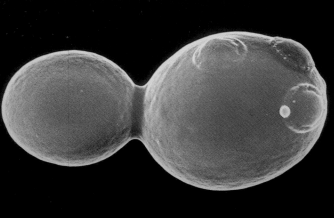

Although most fungi spread by producing spores, fungi also reproduce in other ways. A hypha can break off and grow into a new fungus, completely independent of its parent. This is how most fungi are reproduced in the laboratory. Some fungi, such as the yeasts, live a single-celled existence. They can reproduce by simple cell division, such as an amoeba does, or by another process called *Budding*. When a yeast cell buds, it produces a swelling on its surface that becomes a daughter cell. The parent and daughter cells are connected, but eventually this connection is pinched off and they separate. The new daughter cells soon bud to produce more cells; often they begin to bud even before they separate from the parent. This can lead to the formation of long chains of rapidly growing yeast cells.

A magnified view of baker's yeast.
Inset: The budding cells of brewer's yeast.

An individual species of fungus may reproduce in more than one way. Depending on
conditions, it may produce spores, break up
into individuals, or produce buds at the surface of the hypha. The great reproductive
flexibility of fungi has helped make them one
of the most widely ranging types of organ-
isms on earth. The total number of fungi is
probably between 100,000 and 250,000 species. Most live on land or in fresh water,
although a few live in the ocean.

Mold on a peach.

Different types of fungi may need very different types of environments to survive. Some fungi, such as those found in dung heaps, thrive in temperatures as high as 120 degrees Fahrenheit (49 C). Others, such as snow molds or the molds that attack food in refrigerators, may survive subfreezing temperatures. Light and humidity also affect the growth of fungi. Most fungi, such as mushrooms, mildews, and molds, are found in dark, moist environments. Others, such as lichens, can grow on rocks in some of the driest, harshest climates on earth.

The fungi kingdom is divided into five categories called *Phyla* (singular phylum). There are about 600 different types of fungi in the phylum *Zygomycota*. Black bread mold is a member of this group. So is a very unusual fungus known as the "hat thrower." This fungus grows on horse dung and has evolved a unique method to disperse its spores. It forms tiny clear bulbs that have a black tip, the sporangium. Inside the bulbs, pressure builds up. This pressure becomes so great that the bulb explodes, firing the sporangium up to a yard (.9 m) away.

Covered with a sticky substance, the sporangium sticks to any plant or blade of grass that it hits. When this plant or grass is eaten by an animal, the sporangium dissolves in the animal's digestive tract. The spores pass through unharmed and soon grow into new individuals on the animal's dung, completing the cycle.

Sporangia of the Pilobus *("hat thrower") fungus.*

There are tens of thousands of different species in the phylum *Ascomycota*. Many of them are commercially valuable. Yeasts, for example, are used in baking bread and in brewing beer; truffles are considered delicacies. Other species in this phylum attack commercially valuable plants, causing disease and crop loss in citrus fruits, raspberries, and avocados.

Strawberries engulfed by a fungus.

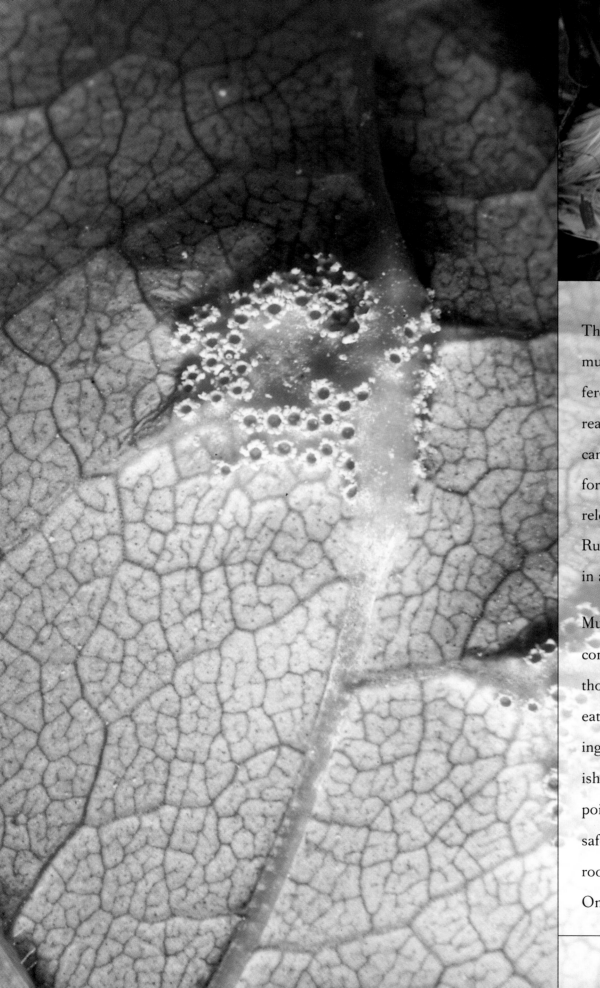

33

The *Basidiomycota* include smuts, rusts, mushrooms, and puffballs. About 25,000 different species are known. Smuts attack cereal grains, grasses, and onions. They often can be seen only at maturity, when blisters form on the plant. These blisters break open, releasing a black powder made of spores. Rust fungi, too, attack crops and can result in a severe loss of productivity.

Mushrooms, on the other hand, are grown commercially throughout the world. Although many types of mushrooms can be eaten, others are extremely poisonous. Eating wild mushrooms is dangerous and foolish; many people die every year from eating poisonous mushrooms they thought were safe. There are thousands of types of mushrooms and some look very much alike. Only experts can properly identify them.

Rust spores on an ash leaf.
Inset: Corn smut.

There are also about 25,000 species of *Deuteromycota*. Penicillin mold is probably the best known of these. In 1929 Scottish biologist Alexander Fleming noticed that chemicals produced by mold killed a wide range of bacteria in the laboratory. Knowing that bacteria are responsible for many diseases, Fleming recognized the potential medical benefit of these mold-produced chemicals. In 1941 other researchers isolated and purified the drug called penicillin, the first in a new class of drugs called *Antibiotics*. These drugs kill bacteria, with few or no side effects in humans. Since their invention, antibiotics made from molds have saved millions of lives throughout the world.

Fungal colonies of Penicillium chrysogenum, *used to produce antibiotics.*

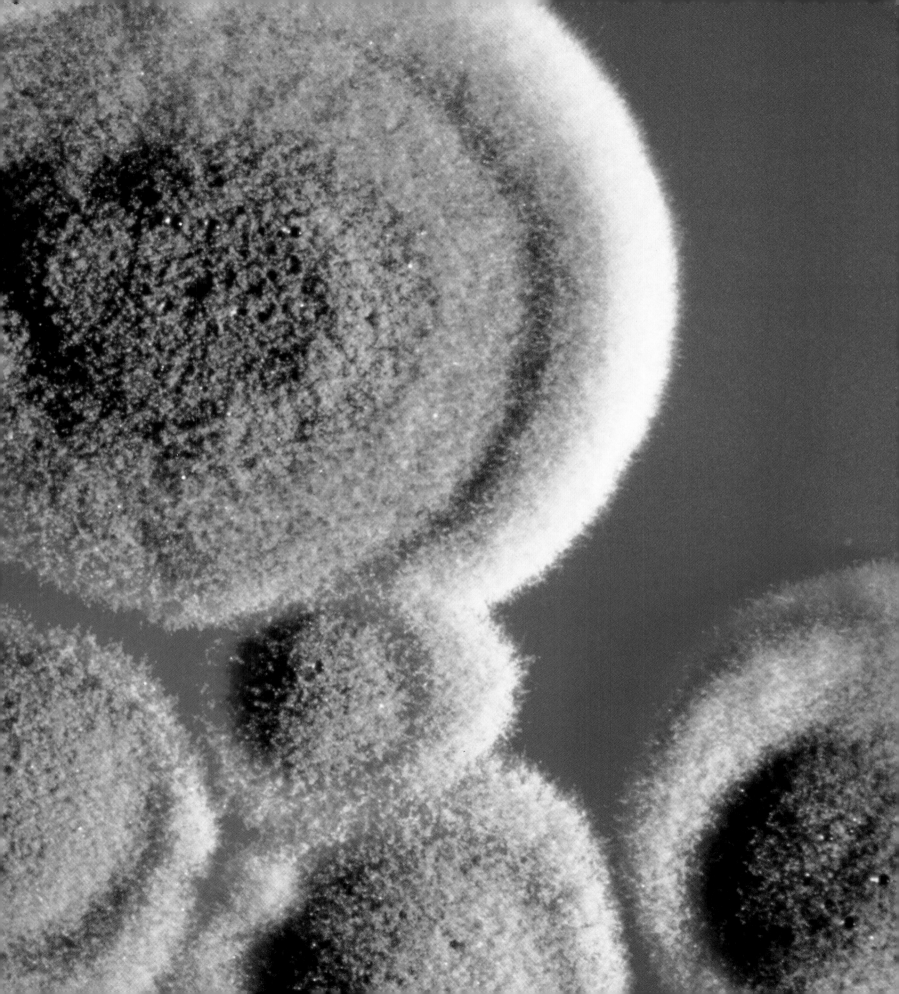

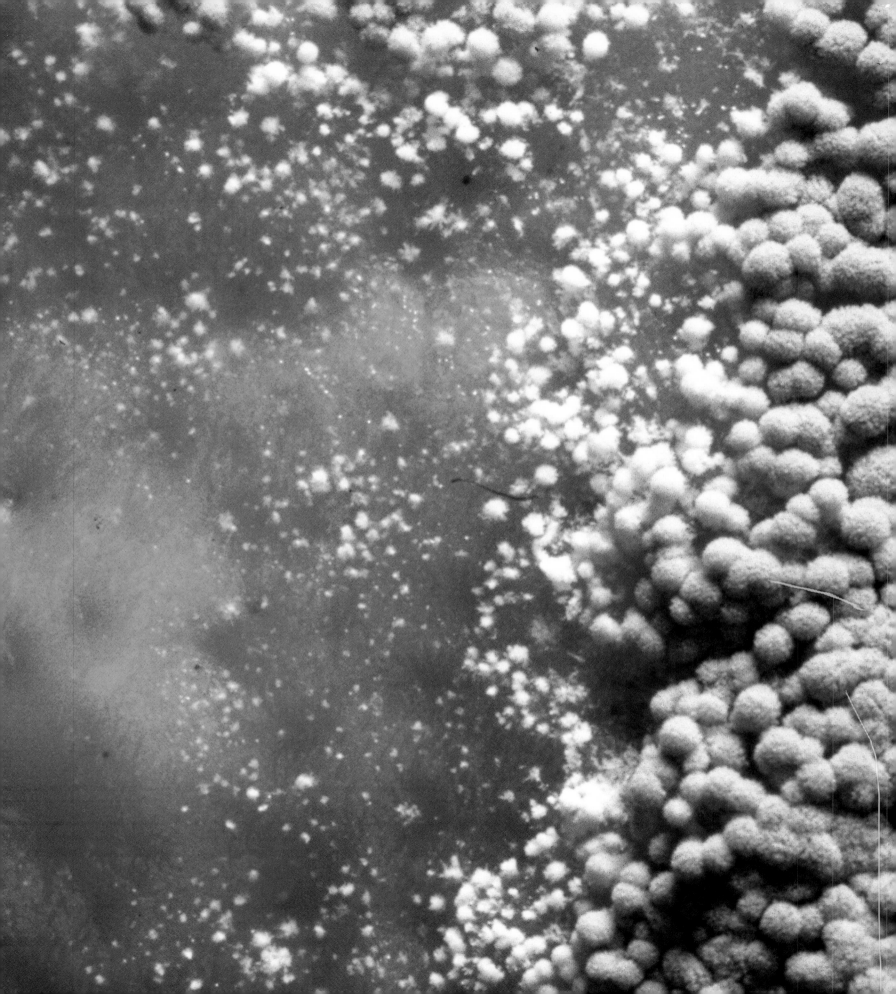

Other molds in this phylum are used to make cheese. After milk is separated into curds and whey, the curds are gathered to be ripened or fermented. Molds, as well as various microorganisms, are used to ripen cheese; many types of cheese owe their flavor and texture to the type of mold used. Roquefort cheese, for example, owes its strong taste and color to the unusual mold *Penicillium roqueforti.*

Mold on a lemon.

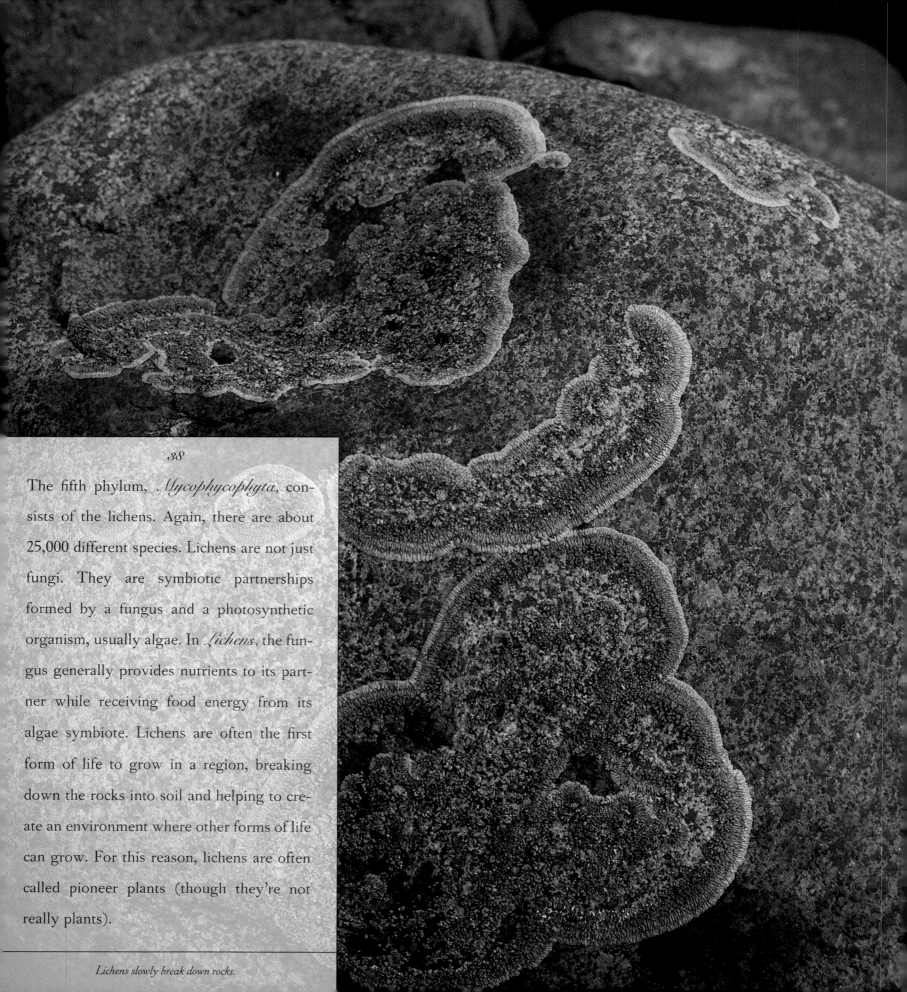

The fifth phylum, *Mycophycophyta*, consists of the lichens. Again, there are about 25,000 different species. Lichens are not just fungi. They are symbiotic partnerships formed by a fungus and a photosynthetic organism, usually algae. In *Lichens*, the fungus generally provides nutrients to its partner while receiving food energy from its algae symbiote. Lichens are often the first form of life to grow in a region, breaking down the rocks into soil and helping to create an environment where other forms of life can grow. For this reason, lichens are often called pioneer plants (though they're not really plants).

Lichens slowly break down rocks.

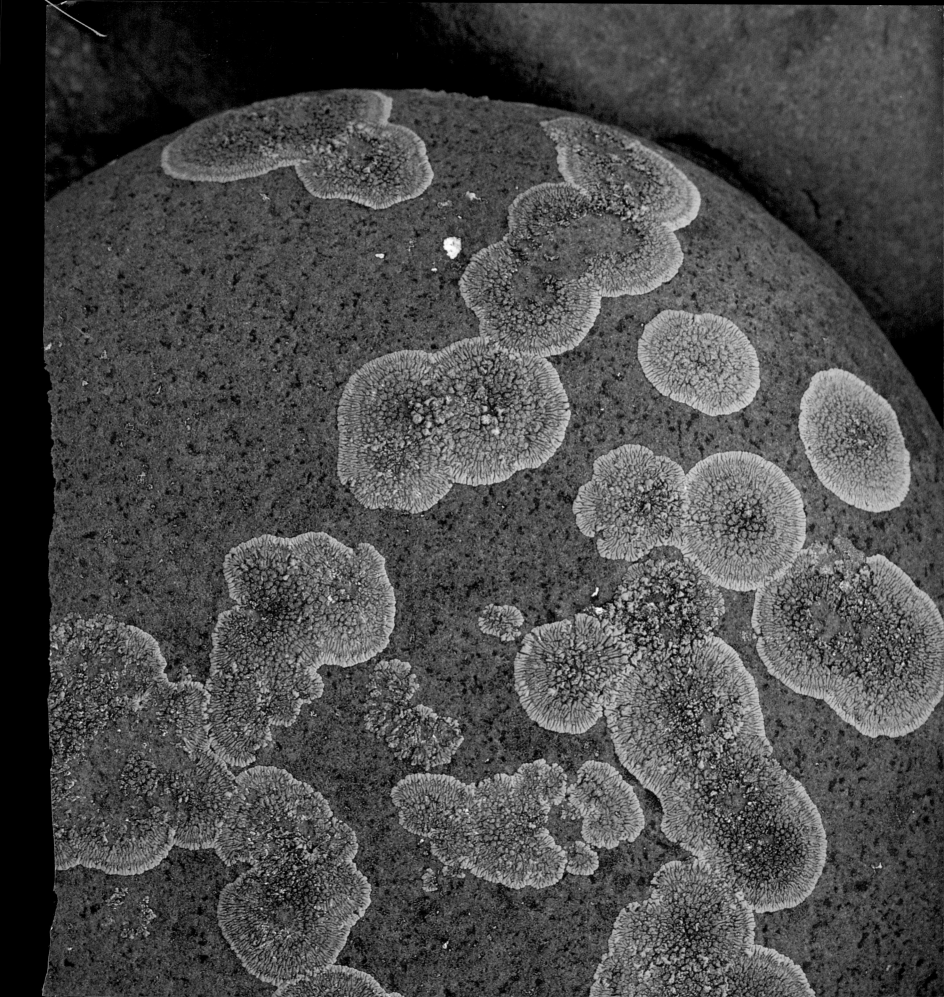

Certainly, *Fungi* are among the most diverse of all life-forms. As we have seen, many of them play a pivotal role in the ecology of our planet. They decompose dead organic material, returning nutrients to the soil; they deliver minerals to tree roots; they are the source of many of our foods and medicines. While moldy leftovers in the refrigerator may still be considered a nuisance, the fungi kingdom as a whole deserves our appreciation.

A lichen garden at Denali National Park, Alaska.